WEST SUSSEX
BARNS AND FARM BUILDINGS

Annabelle F. Hughes.

Following pages
INHOLMES FARM, ALBOURNE (TQ 268169)
A classic example of a boarded field barn in the Weald, with a hipped roof
and a small cattle hovel close by. There are several examples of timber-
framed houses in the village built in the same way,
with similar proportions and profile.

Text   Annabelle Hughes

Photographs   David Johnston

THE DOVECOTE PRESS

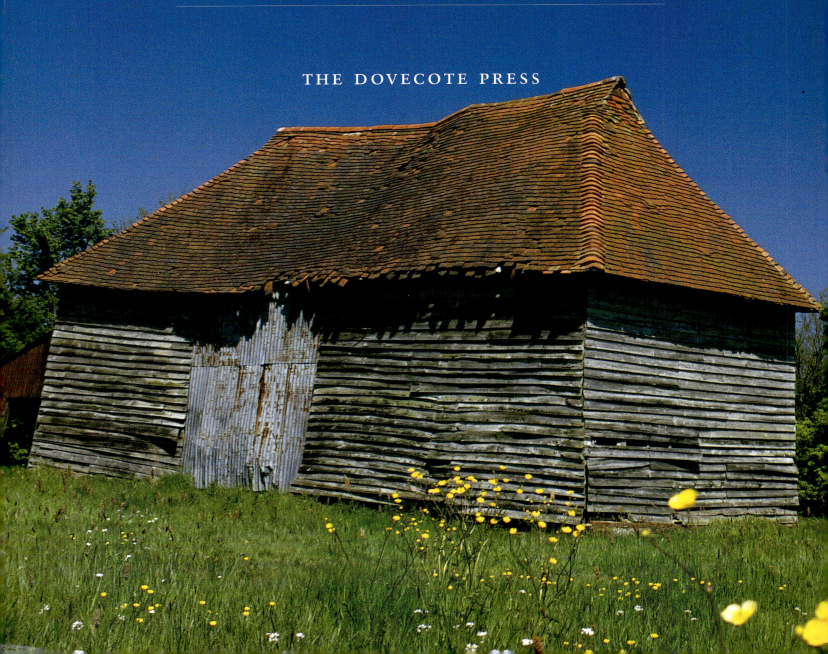

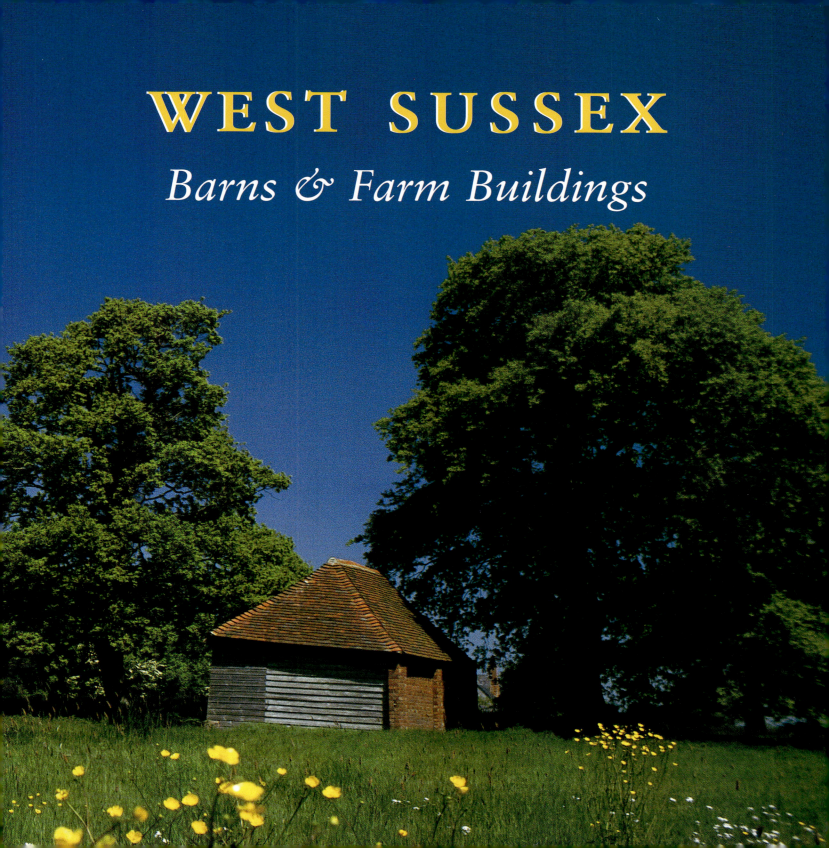

WEST SUSSEX
Barns & Farm Buildings

A downland farmyard near Chanctonbury in the 1930s.
The photograph, by Herbert Felton, is taken from *English Downland*
by H.J. Massingham, published in 1936, and one of the first
books to draw attention to the importance of traditional
barns and farm buildings.

First published in 2002 by The Dovecote Press Ltd
Stanbridge, Wimborne, Dorset BH21 4JD

ISBN 1 904349 00 5

Text © Annabelle Hughes 2002
Photographs © David Johnston 2002

Annabelle Hughes and David Johnston have asserted their rights
under the Copyright, Designs and Patent Act 1988
to be identified as authors of this work

Designed by The Dovecote Press
Typeset in Monotype Sabon
Printed and bound in Singapore

A CIP catalogue record for this book is available
from the British Library

All rights reserved

1 3 5 7 9 8 6 4 2

CONTENTS

THE SETTING 6

CONSTRUCTION 14

BARNS 26
The Coastal Plain
Downland
The Weald

FARM BUILDINGS 85
Granaries
Stables
Other Buildings

DECAY AND RESTORATION 107

Further Reading 116
Photographic Acknowledgements 117
Index 118
List of Subscribers 120

THE SETTING

Until more recent times, farm buildings, like houses, reflected the region in which they were sited. They were traditional, local and functional. They were constructed from the materials that were easiest to obtain at the time, whether it was flint, stone, timber or brick, and were designed to cope with the prevailing weather conditions. Like houses, farm buildings could also reflect both the status of their owners, whether they were individuals or great estates, and the passage of time, for the availability of building materials changed, as did farming practices. If you want to understand the relationship between agricultural buildings and the landscape, it is vital to know something about the county's geology, its geography and its history.

Broadly speaking, West Sussex is geologically and geographically divided into three distinct areas from east to west – the fertile coastal plain, the chalk downland with its loamy soils, and the sandstones and heavy clay of the Weald.

GREEN FARM, WASHINGTON. The roof shapes and proportions of both buildings and haystack blend together almost seamlessly. Rock Mill in the background is a reminder that grain did not have to be taken far for processing. 'Agricultural buildings almost grew out of the landscape'.

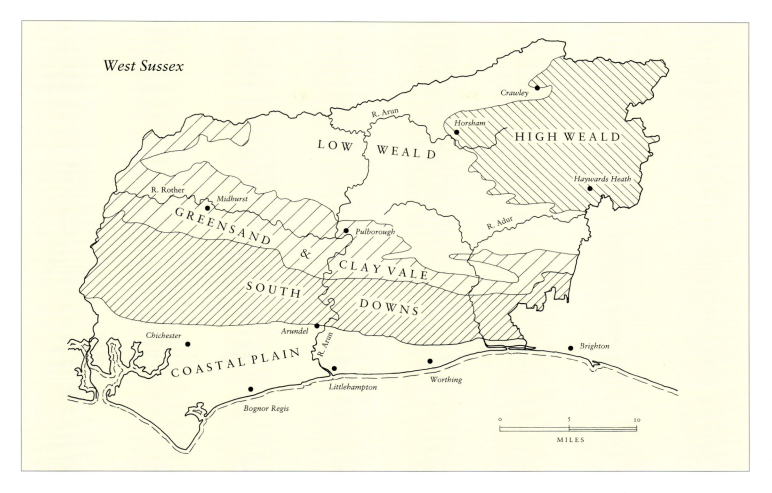

Rising in the north of the county are a myriad of springs and streams that wend their way towards the sea, converging and winding through the chalk downland to emerge as the Adur at Shoreham and the Arun at Littlehampton. Those estuaries, and the natural inlets and harbours around the Selsey peninsula, have been changing their shape over the centuries.

The underlying geology of these three parts has resulted in different soils which have been suitable for different kinds of ground cover. This too has been through many changes, largely brought about by the inter-reaction of nature and man. This inter-reaction was crucial until the Industrial Revolution of the nineteenth century, because

The map illustrates the east-west geological divisions of West Sussex that have influenced agriculture over the centuries.

the economy of the country was overwhelmingly based on agriculture, and the soils dictated the predominant pattern of cultivation and animal husbandry in each area.

Before the agricultural revolution of the eighteenth and nineteenth centuries man was much more dependent on nature's moods. Farmers worked to a longer time-scale, and were more in tune with seasonal changes. Agriculture was labour-intensive, and manpower was plentiful. Agricultural buildings almost grew out of the landscape, constructed of the materials most readily available and

Opposite page & above CASTLE FARM, AMBERLEY (TQ 025132). Typical of the barns that served a manor owned by the bishops of Chichester. In nine bays with two cart entrances, and an aisle to the south, it was built in two phases in the seventeenth century. The nineteenth century additions contain pig-pens (left) and calf sheds (right) and reflect the prosperity of the tenant farmer when it had come under the control of the Arundel estate. This has now been sold for residential conversion.

Basil Strudwick, the last farmer, remembers with affection the old Sussex waggons creaking in from the harvest fields laden with corn sheaves. The waggons would draw into the great barn, where the sheaves were offloaded and stored until threshing time, when the threshing machine would be manoevered into the porch. Inside the north end of the barn is a well (long since capped), where up to the 1930s a pump was fixed, with an engine to pump water to the castle.

most easily worked. Timber was the principal resource, particularly so in those areas where the oak grew like a weed – the clay and sandy soils of the Weald – and up to the sixteenth century the estates on the coast could harvest the timber they needed from woodland they owned to the north. Fewer of these buildings have survived along the coast than in other districts because of modern seaside urban developments.

On the Downs timber could be supplemented and underpinned with flint and chalk, whilst a dearth or scarcity of such natural materials along the coast resulted in a higher proportion of brick being used, particularly from the eighteenth century onwards. The materials used often influenced the profile and proportions of farm buildings, as did changes over time, reflecting the waxing and waning in supply of resources.

Two other crucial elements were social and economic; who had provided the money for the building, and how productive was the farmland it was intended to serve?

One of the basic administrative units was the manor or estate, whose early history goes back before the Norman Conquest of 1066. Originally this was land held by one man or one tribe or family group, and it did not have to be all in one place. In a county like Sussex, where the earliest permanent settlements tended to be established along the coast, it became the custom for each of those settlements to

FULKING (EDBURTON PARISH). This picture, taken in about 1900, shows how closely farming was integrated with village life, the barn (since demolished) then cheek by jowl with village houses. Note the boys playing marbles.

regard areas of downland, wetland meadows bordering rivers, and pieces of woodland as their own, even if they were a number of miles away to the north. These diverse landscapes provided them with different kinds of seasonal grazing and valuable natural resources, most especially timber, which should be seen as the medieval equivalent of plastics – the all-purpose material.

These detached areas or 'outliers' can be identified in later written records of manorial or estate administration. It is important to understand this pattern, for it could sometimes explain early movements of building materials, and a whole range of connections between scattered communities.

For William the Conqueror, Hastings in particular, and Sussex more generally, was his gateway into England. Geologically divided east/west, William was determined to make this gateway secure, and his re-organisation of the Saxon administration built upon and developed north/south corridors, leading into the heart of the country. By the time of the Domesday Survey of 1086, these had become the six Rapes – Arundel with Chichester, Bramber, Lewes, Pevensey and Hastings – under five tenants-in-chief, all of whom had considerable land-holdings elsewhere in England. They were among William's most trusted supporters, and four were directly related to him. Simplistically, each Rape was based on a port, with a principal castle and a major river, and was further divided between tenants and sub-tenants, grouped within 'hundreds'. Arundel and Bramber were approximately the present West Sussex.

The Normans were familiar with the idea of 'manors' and the record of 1086 is organised in manorial units, although many of their names, such as Lavant and Birdham, Amberley and Storrington, Ashington and Ifield, have since become synonymous with towns and villages.

PONDTAIL FARM, SHIPLEY (TQ 162229). A farm on the Burrell estate of Knepp, showing the influence of the planned 'model farm'. Built at the 'tail' of an earlier hammer pond for iron-working.

The traditional view of English agriculture as generally stagnant subsistence farming up to the 1500s, and even then lagging behind until the Agrarian Revolution of the seventeenth and eighteenth centuries, has now been rethought. It has become clear that there were many more regional variations and flexible approaches to land management during the three to four hundred years after the Norman conquest than had been realised. For example, a survey of the Bishop of Chichester's estates made in 1388 gives a picture of very specialised agriculture, showing an appreciation of geographical differences and a high degree of integration between the different manors.

The variations in landscape and soils did mean that productivity in Sussex ranged from the very high, on the coastal plain, to the very low up in the Weald, but from the end of the 1300s, after the ravages of the Black Death, a growing population and the increasing demands of London spurred changes. More areas were brought into cultivation and efforts to improve productivity increased. Landlords found it more profitable to convert to money rent systems and to lease their 'home farms', and so small tenants had more at stake in their land.

As markets for agricultural produce grew, landlords became even more commercial in their approach. They encouraged specialisation such as investment in large flocks of sheep on the Downs, which could be 'folded' on the fields. Although these were seen mainly as manure on the hoof, to improve arable farming, they were also the source of useful by-products. The introduction of a variety of new crops, such as clovers and grasses, helped to invigorate the soils, and feed larger stocks of animals. All these changes increased the need for farm buildings, both to process and store crops and feedstuffs, and to shelter animals. The fertility of the coastal plain and the successful combination of sheep-and-corn husbandry on the Downs made Sussex one of the most advanced agricultural areas of the time.

GRITTENHAM FARM, TILLINGTON. Taken in the 1950s, the photograph shows a traditional farm yard defined by single-storey hovels and stalls.

When Henry VIII dissolved the monasteries towards the end of his reign, he released tracts of land onto the open market, which coincided with the appearance of upwardly-mobile self-made merchants and landowners, looking for opportunities for investment. They did not have as great an impact in West Sussex as in other areas, as the largest ecclesiastical estates, which were based on the coastal plain with far-flung outliers, were not monastic, but belonged to the sees of Chichester and Canterbury. It did stimulate the diocesan authorities towards granting more and longer leases. Smaller holdings, such as those in the Weald belonging to Syon and Westminster Abbeys, and the priories at Hardham, Easebourne and Rusper, were snapped up by families like the Brownes of Cowdray, the Carylls of West Grinstead and Harting, the Gorings of Wiston and Burton and the Bishops at Parham. In many cases the land on such estates was already leased out to ambitious yeomen and gentry.

The sixteenth century saw the beginnings of a huge interest in a more systematic and scientific approach to agriculture. Men like Leonard Mascall of Plumpton, John Norden, and Gervase Markham wrote surveys, discourses and dialogues on agricultural practices, and a great deal of energy was devoted to improving the soil with marl, chalk and lime. In the 1650s turnips were introduced as a fodder crop, making it easier to feed animals through the winter, and 'catch crops' like hemp and flax were grown. Along with these movements to improve went an increase in the loss of the old medieval deer parks, bringing into cultivation land previously devoted to recreation.

On the question of enclosure, which created such hardships and violent protests in other parts of the country, it has been written that 'in this county (it) rarely entailed the wholesale dispossession of the peasantry', but that 'on the contrary Sussex men time and again gave evidence of their ability to effect changes in harmonious co-operation'.

Grain prices soared from the end of the sixteenth century, then fell between 1640 and 1750 by about 25%, but as cattle values increased over the same period, the emphasis shifted towards stock fattening and improving pastures. The more independent small farmers of the Weald prospered, and they invested in 'plant' – barns and cattle-hovels – many of which survived until recently. They had always been used to diversifying in order to survive. The grain producers stepped up production and exported, and then along came the Napoleonic Wars against France, which forced up prices again, rewarding productivity. When the wars ended in 1815, a period of depression was followed by a gradual improvement and rising prices.

From the end of the eighteenth century the larger landlords followed published plans and advice and experimented in building 'model farmsteads', although there are few complete examples in West Sussex. The Stag Park Farm at Petworth is one, and on a much smaller scale Pondtail Farm in West Grinstead, which belonged to the Burrells, whilst gentry like the Shelleys and their successors incorporated some of the new designs for buildings into their estates at Michelgrove (Clapham) and Warnham.

The paintings of John Constable, who died in 1837, have idealised our images of the countryside and agriculture. Taken with the parish tithe maps of the mid-nineteenth century, which were generated by government acts to convert tithes in kind to money rents, they provide a snapshot of how we like to visualise farming before the mechanisation and intensification of the Victorian period got underway. Surviving barns and their dependent structures give us another dimension.

During the last hundred years, two World Wars stimulated production to cope with the need to be self-sufficient, and this accelerated the mechanisation of farming. As farm machinery and agricultural plant has become larger and more complicated many of the traditional buildings are unsuitable for storage on such a large scale, and are no longer needed for processing. Unless alternative uses are found, they are allowed to decay into oblivion, and replaced with utilitarian structures in modern materials that no longer blend into the landscape.

CONSTRUCTION

The barn is the most familiar of agricultural buildings, and just like the traditional timber-framed domestic buildings of Sussex, it was basically a pre-fabricated 'kit' constructed in bays. A bay is best defined as the distance between pairs of principal posts – those posts that extend from ground to eaves and are connected together at the top with a horizontal timber 'tie' – and the distance between those pairs could vary. The divisions of the roof construction were usually dependent on this pattern, and can be identified even in barns with walling of stone or brick.

There are three main types of roof construction – the central purlin with crown-posts, clasped side-purlins with queen or raking struts, and butt-purlins in line or staggered

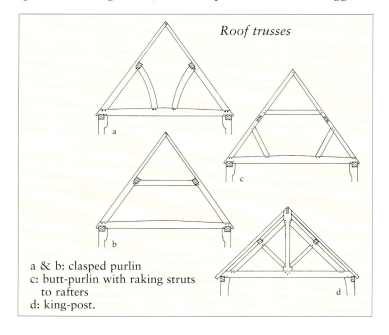

Roof trusses

a & b: clasped purlin
c: butt-purlin with raking struts to rafters
d: king-post.

UPWALTHAM MANOR FARM, UPWALTHAM (SU 962138). This interior illustrates a side-purlin roof with raking struts from the ties and wind bracing in the planes of the roof.

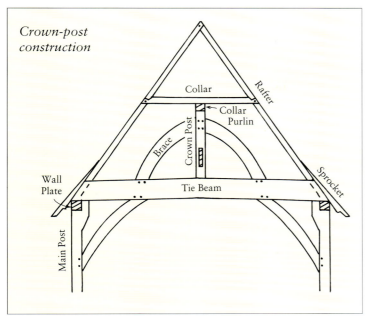

Crown-post construction

– and earlier roofs tended to be hipped at both ends. Chronologically they fall into this sequence, with crown-posting being the general rule up to the middle of the sixteenth century, to be supplanted by side-purlins and finally by butt-purlins, which could make use of shorter lengths of timber.

There are a few barns with all or some of their crown-posts left, mainly up in the Weald, and they are something to treasure. It is more common to find barns still with the large-panelled, heavily braced wall-framing that went with crown-posting, but whose roof has been reconstructed in a later form. This could have been done when thatch was changed to tiles, as part of a repair, or as an adaptation

SENNOTTS FARM, SCAYNES HILL, LINDFIELD (TQ 391235). This is one of the rare surviving examples of crown-post roof construction that is characteristic of early timber-framed barns. The crown-post is braced in only two directions, to purlin and tie, and the rafters have retained their collars.

SULLINGTON MANOR FARM, SULLINGTON (TQ 098132). An example of the carpenter's marks found in many timber-framed buildings, and were part of a numbering scheme to help the assembly of the pre-fabricated 'kit'.

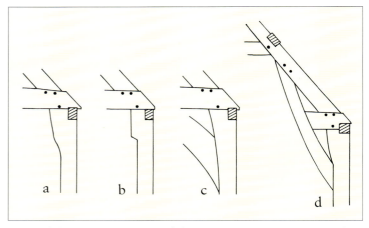

Some of the variations at one of the most important joints in timber-framed buildings – the post-tie-eaves plate. A, b & c show some of the different jowl profiles, while d illustrates the 'sling-brace', an 18/19th century method of improving storage space.

when crown-posts, collars and the central purlin got in the way of storage or machinery.

The majority of surviving traditional barns have some variation of side-purlin construction, where the roofs generally terminate in gables or half-hips. The planes of such roofs are often strengthened with wind-bracing between principal rafters and purlins, which was designed to prevent the rafters from 'racking' or leaning towards one end, and forms a pleasing pattern of curved timbers.

The size of a barn will vary according to the purpose for which it was built, to the acreage of the farm to which it was attached, and to the changes in requirements over time. A 'threshing barn' had a central bay with large doors at each side, creating a through draught that helped separate the wheat from the chaff during the winnowing process. This central area was flanked by one or two further bays where hay could be stored. Sometimes evidence remains for the paved or boarded 'threshing floor' in the central bay, and the slots in the posts into which vertical boards were dropped to prevent the straw and chaff spilling into the rest of the barn. The high doors on one side enabled loaded wagons to be driven in, and out by the lower doors on the opposite side when they were

BUTTERBOX FARM, SCAYNES HILL, LINDFIELD (TQ 383238). The great sweep of a hipped roof down over end and side aisles. In 1581 Richard Coleman 'for his land Butter Box' supplied 4 feet of churchyard fencing to Lindfield church.

The original part of the farmhouse is a medieval hall-house, which has since been added to and altered. During the Tudor period a beam from the house was incorporated into the barn, only to be finally retrieved for re-use in an upper room.

emptied. If a barn has doors on only one side, it was intended for other uses.

Empty mortices in posts, ties and rails, and original window mullions at first floor level can be evidence that one of the end bays of a barn had a first floor. This could provide an integral granary at first floor, with grain bins, or even basic accomodation for farm labourers, and was

17

CASTLE FARM, AMBERLEY (TQ 025132). The jowled principal post is braced to the arcade plate of the aisle (top left) and tie (centre right) and there are curved wind braces in the top right corner.

WISTON GREAT BARN, WISTON (TQ 148123). Built on the Goring estate, this is a rare example of chalk 'clunch' blocks used with brick dressings to the ventilation slots.

Opposite page top BURPHAM FARM, BURPHAM (TQ 0430890). A thatched barn with a cattle hovel; now converted.

Up until the first quarter of the twentieth century, vast flocks of sheep were annually brought down off the surrounding downs to be dipped in a specially constructed dip built into the stream below the farm. Such were the numbers, and the dust, that every door was closed, including that of the church, which was normally left open both day and night.

Opposite page below LAGNESS FARM, PAGHAM, illustrating the combination of upper level boarding, affording good ventilation, with ground storey stonework. Many barns were originally thatched, using a by-product of the harvest. At its most functional, thatching was a skill to which many local men would turn.

sometimes intended from the beginning, or could be a later adaptation. In the same way, it is uncommon to find ridge boards used along the apex of the rafters until the middle of the eighteenth century, and they are sometimes a clue to a rebuilt roof over an earlier barn. The timbers of many barns are evidence of the 'mix-and-match' approach, when the best of several barns would be recycled into a 'new' construction.

The side walling of barns is another area which has

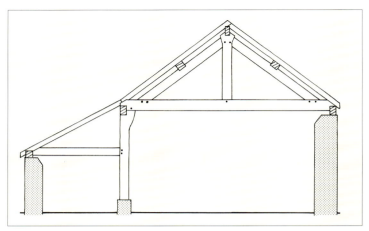

Aisled construction, showing the aisle to a solid walled building with a king-post roof.

MARSHALLS FARM, KIRDFORD (TQ 012246). An example of the flexibility of barns; partial flooring has been built into one end and modern grain bins installed.

THE OLD BARN, OLD PLACE, PULBOROUGH. A drawing in the early nineteenth century showed this building with a thatched roof, and in spite of the elaborate stonework of the doorways and windows, it was part of the L-shaped 'home farm' buildings, surrounding the courtyard of that part of the manor of Pulborough which was anciently held by the Hussey family. It would have included stables, possibly lodgings for farm workers, a granary and barn, and was designed as evidence of the landed wealth of the owners.

Built with a crown-posted roof, possibly by the de l'Isle family, tenants in the fifteenth century, by 1953 it was reduced and converted to a dwelling. When the tenant farmer, John Boxall, retired in 1902, all his farming equipment was sold at auction, including three teams of four horses, a bench on which pigs were killed and dressed, a lark net and eel trap.

UPPER HORTON FARM, UPPER BEEDING (TQ 209113). Although the corrugated sheeting has little charm, it has kept the barn weather-proof and secured its future.

Upper Beeding was the birthplace of Michael Blann (1843-1934), a shepherd who collected Sussex folk songs, thus helping preserve them. His tin whistle and song book are now in Worthing Museum.

often undergone a series of changes over the decades. If the framing is examined closely, it will often reveal the notches and runnels for the staves of earlier woven wattles, which would be partly or wholly daubed with a mixture of puddled mud, chopped straw and dung. Many barns were always boarded, sometimes with vertical planks, although obviously the older the building the rarer it is to find boarding that is original. Sometimes there is evidence for a mixture of wall-covering – boarding to the lower level, with wattle above, which could be daubed or left open.

Many barns are aisled, the roof carrying down beyond the side or end walling to lower aisle walls. Aisles could serve a number of functions – from storage for a whole variety of items to animal stalls. They could be added or taken away, as barns were supremely flexible, with

LOWER CHANCTON FARM. A good complex of farm house and associated buildings on the Wiston estate, where the mixed farming typical of the Weald prevailed. Chanctonbury Ring in the background was planted by one of the Goring family in the 18th century; it was replanted by his descendant after the devastations of the Great Storm of 1987.

potential for all kinds of change and adaptation to suit the prevailing economic climate – for grain when prices were high to stock and fodder when cattle were in fashion.

When barns or other farm buildings are built from flint, it is well-nigh impossible to construct right angles, so the corners and openings are generally edged or 'dressed' with brick or stone. Flint had several advantages. It was readily available on the chalkland of the Downs, could be picked from the fields with cheap local labour, and once built, the structure needed less maintenance. For a variety of similar reasons farm buildings were constructed partly or wholly of brick from the 1700s. As time went on timber was less easily available in the downland and coastal areas, and farmers or landowners amassed the funds and means to provide more permanent, maintenance-free buildings.

Apart from the evidence of the buildings themselves, there are some surviving historical maps and written records. From the evidence of manorial accounts and surveys, it seems likely that most barns were originally thatched, while those of higher status often had sandstone tiles, quarried in the Weald.

A survey of the manor of Bury (near Bignor) in about 1589 contains the following description of the farm buildings of the demesne or 'home farm' which was close to the church:

SADDLESCOMBE FARM, NEWTIMBER (TQ 278115). Part of the hamlet built up by the Petworth estate around a farm that belonged to the Knights Templars until 1308.

'One Great Barn for Corn one Hay Barn One Grainery one dove house and stable to stall Oxen in all which houses be covered with Horsham Stone and with tile'.

Most thatch and stone has been replaced with the clay tiles that were used on later builds, and more recently with the ubiquitous corrugated iron and asbestos sheeting.

Many larger barns are mistakenly called 'tithe barns'. In return for maintaining the chancel of the church and providing for worship, the rector appointed to the living of a parish was entitled to the 'tithes' or tenths of all the local produce. Over time, many parishes were taken over by monasteries, colleges or even wealthy lay people, who were therefore in the position of a rector. They kept the great tithes of corn and hay, and used the smaller tithes to pay a vicar to serve the parish in their place. Only a barn which

CHURCH FARM, WASHINGTON (TQ 119128). Its position suggests this could have been a genuine 'tithe' barn.

The roar of traffic on the nearby A24 is in marked contrast to Gerrard Washington's childhood memoir of holidays spent at the farm before the First World War: 'From the window overlooking the flower garden we lean out to smell the evening primroses – we are content to lean on the sill, listening to the horses stamping their hooves on the stable floor, and the cows occasionally lowing.'

was built by a rector (individual or corporate) is truly a tithe barn.

Some of the most impressive barns that survive in England were built by the owners of large estates, who needed the space to process and store huge quantities of grain and hay, and before the Dissolution of the Monasteries under Henry VIII, many such estates were owned by monks. This is less true of West Sussex, where generally speaking, the large barns which survive are from the second half of the sixteenth century, and can be found on farms of the estates that belonged to the bishops of Chichester, the archbishops of Canterbury, the owners of Arundel Castle and Petworth, or a few wealthy gentry families (some of whom remained Catholic) who bought up the smaller monastic estates.

For centuries West Sussex has been dominated by the Arundel estate of the FitzAlan and Howard Dukes of Norfolk, the Cowdray estate based at Midhurst, the Petworth estate originally belonging to the Percy family but now held by the Leconfields, and the Goodwood estate of the Dukes of Richmond, which was developed from the beginning of the eighteenth century.

For examples of their influence one has only to look at the complex of buildings at Saddlescombe, north of Brighton, partly developed by the Brownes of Cowdray in the late sixteenth century, and much from the late eighteenth and early nineteenth century when it was tenanted from Petworth, to the distinctive buttressed brick barns on farms like Bury, Amberley, Clapham and Warningcamp that were part of the Duke of Norfolk's estate in the nineteenth century, and to the flint constructions on the Goodwood estate.

BARNS

Farm buildings that survive from the fifteenth or sixteenth centuries must be some of the best examples, as they have lasted 500 years without being replaced, although they have probably been repaired and modified. It is almost exclusively barns that have survived from this period, and not many of them, but as they were built by wealthy landlords, they are not typical of the local agricultural scene. Until the late eighteenth century the formal cluster of buildings set squarely around a yard was not the norm, and although it is possible to find a traditionally built house with a barn close by, in many cases farm buildings were sited close to tracks on the edge of farmland. We know that there were farm buildings other than barns, but few have survived from before the late 1700s. However, it is likely that their range and construction was similar to those that replaced them.

Agriculture in West Sussex fell into two distinct groups; the grain producing areas of the coastal plain and chalk downland, with sheep as a vital part of the 'folding' system, and the cattle-producing, dairying areas of the Weald, so the traditional farm buildings in each area reflected different requirements. There were barns and granaries close to homesteads for grain processing and storage, remote barns and open-sided hovels for ploughing oxen and the management and care of the large sheep flocks, field barns in the Weald for cattle shelters and feed. Against this overall pattern must be set the small subsistence farmers who practised much more mixed farming in both areas, and whose buildings have long vanished.

The records of the large estates, many of which belonged to the Church, are those that survive best and it is among these that we find the earliest references to farm buildings:

'John le Honte . . . shall receive timber by the view and livery of the monks for the repair and other needs of the buildings of the tenement, provided the dilapidations do not arise through the action of the said John or through or his men's lack of care . . . '.

John le Honte was a tenant up in the Weald near Horsham, but he held his land from Sele Priory, near Bramber Castle. These details of his lease were written in 1285, and suggest several timber-framed buildings, one of which must have been a barn. From about the same year there are these instructions for tenants of the Slindon estate of the Archbishop of Canterbury:

'. . . if the granges are fallen down, (Martin Anonentun) with the other tenants . . . in the vill of Slindon and . . . in the Weald, when timber has been carried to the site and squared up shall do all the carpentry excepting partition walls and doors, and for doing this the lord shall provide the master carpenter with 1 seam of wheat, 1 bacon carcass and a cheese of 1 pound for his expenses'.

The word 'grange' is one that often appears in these early records, and describes what we would now call a barn. This extract shows how the landlord of a large and scattered estate – in this case a prince of the Church – could utilise both natural and human resources.

From similar records for the estates of the bishops of Chichester in the thirteenth century, Amberley tenants are directed to make and mend a 'barn, byre and hayloft . . . if needed' under the supervision of a master carpenter with wood coming from the northerly landholdings near Rudgwick. A group of tenants from the Ferring estate were given detailed instructions, and were to be fed for the duration of the work:

'All these (four named men) shall come to Ferring with carpenters' tools to make a haybarn, if needed, or to mend it, and they shall help the master carpenters to flaw the timber and carry it . . . but shall make no holes to put the posts in; they shall find the timber for the said barn from their own wood at the lord's will. They shall have the bark and the lop and top, and the lord shall throw the timber. If the master carpenter is not content with their work they shall make fine at the lord's will and return, and they shall stay until the barn is built.'

As well as these barns, there were references to granaries, stables and byres on the episcopal estates. When the lands of the Knights Templar were seized and listed in 1308, among their assets at Saddlescombe (north of Brighton) and Shipley (in the Weald) were an ox-shed and a cow-shed as well as two granges.

Dovecots, chicken sheds and sheep sheds could be found on the estates of aristocratic landlords like the Percies of Petworth, and the de Braose family of Wiston lands (near Steyning). However, throughout these records it is the 'grange' that appears most frequently, for storage of all manner of crops, but in particular corn, hay and barley.

A survey made for the manor of Bury near Pulborough over three hundred years later, in 1669, which then belonged to the Duke of Norfolk, contained similar instructions, this time with some interesting further details:

'They are to build the Lord's Barne at Bury with the Lord's timber and are to have all the Chucks and Chipps which come off and are remayning of the timber when the work is finished The Lord is to find the Master Carpenter And the said tenants the rest and are to build all the Barne to the Roofe or Covering And they are to have the timber of the old Barne provided the Tenant to the said Farme or the Lord's Bayliffe shall not employ it about any other Building And the Lord is to give the Tenants at the building the said Barne One quarter of wheate one Bacon hogge One cheese and Six pence in money And likewise the said Tenants are to build the Lord's Stall at the said Farme of Bury.'

Here is documentary support for the use of an 'estate carpenter', for rebuilding at a time when agricultural improvements were well underway, for recycling building material and for a new 'stall', possibly to accommodate the increasing numbers of stock that were being fattened indoors rather than grazed.

The pleasing groups of farm buildings that we have been accustomed to are the result of generations of expansion and contraction in response to boom and depression, often contain wide variations in age and style, and are usually focussed on one or two barns. Where there are formal arrangements of ranges of buildings mostly of one period, they are likely to date from the end of the eighteenth century and the first half of the nineteenth, and to be the products of the 'model' farm era.

The Coastal Plain

CHURCH FARM, SIDLESHAM (SZ 856988). A fine timber boarded barn on stone footings, although its thatched roof is badly in need of repair. A rare survival on the coast, probably explained by the fact that it was on a manor belonging to the bishops of Chichester, and the income from the church supported a canon of the cathedral. It is aisled around all sides, but the aisling has to be set back at each side of the central doors. When a smallholder of the parish died in 1620, he had 6 acres sown with wheat, barley and tares, and 2 pigs.

Above BINSTEAD NURSERY FARM, WALBERTON (SU 985055). This fine flint barn dressed with brick is characteristic of the coastal plain. The half-hipped roof indicates it was built in the seventeenth or eighteenth centuries.

Left LITTLE OLDWICK FARM, LAVANT (SU 846076). A beautiful example of random flintwork, with brick dressings for the ventilation opening.

BONHAMS FARM, YAPTON (SU 974035). A stone barn with a sweeping roof, half-hipped, with additions to each end. In earlier times, Yapton was held by the Zouche family from the Percies of Petworth.

Downland

MANOR FARM, BURY (TQ 008131). The arches beneath this buttressed building, built in 1879, show that it is less of a barn and more of a cart-shed with storage above, possibly for grain. Very grand as befits a tenancy of the Dukes of Norfolk at Arundel.

Manor Farm was the workplace of one of Sussex's longest-serving agricultural labourers, Charles Kilhams (1861-1939), who started there aged seven, and was still at work when he died seventy-one years later.

Two views of SOUTHVIEW FARM, BURY (TQ 009139). Known locally as Bone Barn, the barn is by Bone Field and tradition has it that it was associated with a body-snatcher.

Writing in 1948, Lilian Brown tells how 'after rifling a grave [he] brought his spoil – later to be sold for purposes of vivisection – over the stone stile on the north side of the churchyard and then by quiet tracks to the stream; and so over the main road to the place where he had his sheds – a field thereafter known as Bone Field'.

The interior shows that the building included a 'wany' tie and is of slighter timber scantling, typical of utilitarian constructions from the mid-seventeenth century onwards.

CLAPHAM MANOR FARM, CLAPHAM (TQ 095066). The barn's boarding is in a sad state of repair, in contrast with the brick stable block.

The parish was once dominated by the great house of Michelgrove, held for generations by the Shelley family, which was demolished after 1828 when it was purchased by the 12th Duke of Norfolk. The remains of a dovecote, with a clock-tower storey built on later, stands on a hill nearby.

From 1768 to 1814, three successive members of the Tompkins family, land agents to the Shelleys, kept a notebook that has survived. From 1768 two men were paid a guinea a year to kill rats, and over 8 dozen pigeons were sent to London.

Above BESLEY FARM, COLDWALTHAM (TQ 015156). A fine stone barn with an unusual brick arched entry. The single-storey stock shed has been built in line, rather than at right angles, which is more typical.

Left LONGFURLONG BARN, CLAPHAM (TQ 098077). A brick buttressed construction typical of the nineteenth century. In 1804, the Tompkins diary notes 'Built the New barn at Poling with Sixteen Brick buttresses'.

WATERGATE HOUSE FARM, COMPTON (SU 785119). A fine boarded barn on stone foundations in excellent repair. It is very likely that it was used for several purposes.

SMOKEY HOUSE BARN, GRAFFHAM (SU 928204). Well-built in stone, the low doors can be accommodated without setting the aisles back.

LEGGES BARN, HEYSHOTT (SU 897179). A stone barn with brick dressings, forming a yard with the stables. Here the high doors have been accommodated in the aisle by forming a 'porch' with a hipped roof.

Four views of SADDLESCOMBE FARM, NEWTIMBER (TQ 278115).

Opposite page top The approach to the fine collection of flint and brick buildings that make up this complex, now owned by the National Trust. The farm was among the possessions of the Knights Templar in the twelfth century. When an inventory of their possessions was taken in 1308, before they were disbanded, they were cultivating 163 acres here, had two barns, an ox shed and a stable, with a windmill worth 13s 4d, and ran a flock of 600 sheep. The Robinson family who farmed here from 1853 had a flock of 900 animals. A rare example of a 'donkey wheel' survives, constructed to make it easier to draw water from the well, which was deepened by 10ft to 160ft after 1853.

Opposite page below One of the barns with a slate roof; the walls are of chalk blocks cased with flints. Some of the buildings may have been constructed by the Brownes of Cowdray at Midhurst, who acquired the farm immediately after the Dissolution of the Monasteries.

Above Saddlescombe in about 1900. Note the flock of sheep in the lane.

Right Detail of the flint work with remains of a brick dressed vent.

WICKHURST BARN, POYNINGS (TQ 254116). Tucked behind the downs, this superb flint and brick barn stands isolated on the edge of farmland, the doors opening straight onto the present roadway.

Opposite page WARREN BARN, SLINDON (SU 965112). The gable end of this barn shows ingenious use of brick banding in a squared pattern to strengthen the flintwork. The boarding above may be evidence of a change from half-hip to gable. The doors are accommodated with a high porch with a pitched roof. Most of Slindon was a manor belonging to the Archbishops of Canterbury. Sadly, the barn was gutted by fire in 2000 and has recently been repaired by the National Trust.

Oppoiste page SOUTH STOKE FARM, SOUTH STOKE (TQ 024099). Another buttressed construction indicative of the nineteenth century and the hand of Arundel. The barn is set against a bank so that a stepped platform has been built to get to the door.

Below TURNPIKE FARM, STOPHAM (TQ 025185). This is built in local ironstone with its characteristic colouring, and with its open end seems more like an implement store than a conventional barn.

For centuries Stopham parish has been associated with the Bartelot family, and boasts one of the finest fourteenth century bridges in the county.

43

MANOR FARM, WALDERTON (STOUGHTON) (SU 801114). Although the roof is not picturesque, it is extremely practical and serves the purpose of protecting the barn, which has now been converted to a private residence. As the name suggests, this would have been the principal farm of the parish.

MANOR FARM, SULLINGTON (TQ 098132). This fine aisled barn, in stone and timber with a high porch for the doors is close to the church and farmhouse, both with elements dating from the thirteenth century and earlier. This is probably one of two barns mentioned in 1635, 'wherein is one stable and one dovehouse', and which was the only one surviving 'in tolerable repair' in 1724.

Mrs Pasmore of Coombes remembered dances being held there in the 1920s, although now it is used principally to store corn. The Coverts, who were lords of the manor for many generations, had a hunting park and lodge at Broadbridge Heath, near Horsham. The recumbent stone effigy of a knight in the church may be Baldwin de Covert who paid local tax in 1378.

Opposite page THE GREAT BARN, MANOR FARM, SULLINGTON (TQ 098132). The magnificent interior of the barn, with its braced ties and raking struts.

Right THE GREAT BARN, MANOR FARM, SULLINGTON. This redundant window suggests that part of the building was once floored, possibly to accommodate farm-workers.

Below MANOR FARM, SULLINGTON. A typical open-sided hovel onto a cattle yard.

DIDLING MANOR FARM, TREYFORD (SU 838186). A wonderfully haphazard cluster of buildings of different ages.

Left DIDLING MANOR FARM, TREYFORD (SU 838186). Two barns in a variety of materials. The brick detail in the flint work suggests that this barn included a dovecote.

Below BLAKEHURST FARM, WARNINGCAMP (TQ 045067). Yet another of the fine buttressed barns on part of the Arundel estate, this time in flint, with brick details, most notably on the gable end. Eight bays long, it rivals the framed barn at Amberley and also needs two entrances.

When William, the son of Warin and Eloise of Blakehurst, sold a couple of acres here in 1263, it was noted that his parents had been giving a 'peppercorn' rent. By 1563 the Palmers of Parham had the tenancy.

EAST CLAYTON FARM, WASHINGTON (TQ 109138). Here the single buttress, timber shoring and brick patching are all clear signs of on-going repairs. The brick 'string course' in the stone work echoes details that can be seen on domestic buildings.

The Weald

Opposite page THE GRANARY, GARSTONS FARM, BOLNEY (TQ 258219). This fine timber-framed granary, with brick infilling, dates from the late sixteenth or early seventeenth century, although it has a wonderful studded door (*Right*), with a date of 1471 scratched upon it, which is thought to still swing on its original hinges. This is a cautionary note, as dates were rarely placed on traditional buildings before the late sixteenth century, and even then, they often only indicate additions or alterations.

The earliest parts of the farmhouse can be dated to about the middle of the fourteenth century, and the Garston family were living in the parish from at least 1308 to 1525.

Above THE GRANARY, GARSTONS FARM, BOLNEY (TQ 258219), showing the use of a blacksmith's 'strap' used to repair framework that has begun to pull apart.

RYELANDS FARM, ARDINGLY (BALCOMBE) (TQ 321281). John ate Ree paid tax here in 1327, and it is likely that the farm house dates from that time. This later stock barn still retains the hay rack for feeding cattle when they were being 'wintered' or fattened inside.

ROWNER FARM, BILLINGSHURST (TQ 071269). Making use of stone available in the area, the single-storey stable range abuts a barn with half-hip roof. The Rowner family who gave their name to this farm were living in the parish in the early fifteenth century, and with its water mill it was part of the Goring family estate in 1886. At auction that year the farm buildings are described as: *Spacious Timber and Tiled Barn and range of Sheds, Chaise-house, Stabling for four Horses, Stock Yard, partly walled in; Second Yard, Fatting Stall for Ten Beasts, with Open Sheds adjoining, and a Large Dutch barn.*

NYES HILL FARM, BOLNEY (TQ 243219). A fully framed and boarded stock shed set at right angles to a large barn with half-hipped roof to form a yard. Both may date from the end of the seventeenth or early eighteenth century.

POSSESSION HOUSE FARM, ITCHINGFIELD (TQ 116269). Sadly in a state of great decay, in spite of evidence that it is still in use. It is probably only surviving because of the substantial lower storey of brickwork, with ornamental vents. There are many romantic legends attached to this farm, not least because it is where the notorious Rapley smuggling gang were finally rounded up in the early nineteenth century. Known as Wares, Withals, and Coombeland in 1615, and simply Wares on the 'church panels' list of 1707, the farm was sold away from the local Muntham estate and later bought back into 'possession' again. The name change appears in 1731, and at times has been contracted to 'Session House', giving rise to wild theories about courts and Judge Jeffreys!

Above MARSHALLS FARM, KIRDFORD (TQ 012246). A fine aisled barn with roofed porch over the doors that probably opened onto the threshing floor. The name 'marshall' originally described a man in charge of horses, and later to a smith, and a Thomas 'Mareschal' paid tax in the vicinity in 1296. The Downer family farmed here for nearly 200 years until the early twentieth century, as tenants of the Petworth estate.

CRIMBORNE HOUSE STUD FARM, KIRDFORD (TQ 027236). This building looks much like a stable with hay loft or granary above. In 1798, when William Knight tenanted the farm from Bedham manor, he had 10 cattle, 9 sheep, 4 pigs, 5 draught horses, 2 wagons, 2 carts, 5 quarters of wheat and oats and three loads of hay and straw. Between 1779 and 1834 sales of oak, beech and other timber brought in almost as much per acre as the farm rent in the 1950s.

Above and opposite page OLD ELKHAM FARM, EBERNOE (SU 993256). Apparently a field barn and hovel in random sized stone blocks, with side-purlin roof, raking struts and wind braces to the half-hipped roof, this was once part of a larger group of buildings.

The farmhouse was built in 1565 by Robert Ffyst, who died before he could roof it, and was finally demolished in about 1960. Although Ebernoe now has its own church, the area once lay within the parishes of Kirdford and Northchapel, and still retains much ancient woodland with a magical atmosphere. In the early seventeenth century Elkham came under the Wiston estate, then in Sherley ownership, but in the 1300s John Elkham's land was a tenancy of Pallingham manor, held by the FitzAlans of Arundel.

Left and opposite page FRITHFOLD FARM, EBERNOE, NORTHCHAPEL (SU 981287). The farm name is derived from the Saxon phrase for an animal enclosure (fold) in the woodland (frith). In 1675 Thomas Cox of Frithfold died leaving the following stock and crops. He was representative of a comparatively substantial farmer of the period:

	£ s d
fourteen acres of wheat at seventeen shillings the acre comes to	11.17.10
five acres of peas	1.10.00
eighteen acres of oats	10.10.00
five pairs of oxen	16.00.00
a pair of steers	7.00.00
five cows and three calves	14.00.00
three yearlings	4.10.00
one horse beast	4.00.00
four hogs	3.00.00
four sheep	1.06.06
one waggon	4.00.00
one dung pot	1.00.00
one ox harrow two horse harrows two ploughs three yokes three chains and some harness	1.10.00
hay in the barns	2.06.06
eight loads of stones	2.00.00
for making of two thousand faggots	2.05.00
two hundred of hop poles for the cutting of them	2.00

Oppsite page GODSDENHEATH FARM, LODSWORTH (SU 921221). A barn and hovel close to the farmhouse; the barn has the half-hips characteristic from the seventeenth century onwards.

Above WOOLHURST FARM, LODSWORTH (SU 925231). An aisled barn in a variety of materials that suggest a long sequence of extension and repair, this was once part of a fine, unspoilt cluster of old buildings. Some of the buildings have been demolished and others converted to a variety of uses.

Following pages WESTLANDS FARM, PETWORTH (SU 999233). A fine group in the landscape. Algie Moss, who farmed Westlands until his retirement in 1974, was a staunch traditionalist, and almost the last of a dying breed. He hand-milked his herd, and cut his corn with a horse-drawn binder, carrying it home in old Sussex waggons.

GOFFS FARM, NORTHCHAPEL (SU951284). This was a copyhold called Steers and Cumbers in 1610, when John Magicke was living here. The old farmhouse was encased with brick and 'modernised' in 1657, by William Yaldwin, who put the date and his initials over the porch. A classic example of 'upward mobility', he was an ironmaster at Frith in 1636 and high sheriff of Sussex in 1656.

The barn would probably have been a hundred years old in 1779, when William Taylor farmed 34 acres of 'Magicks'. It was not called Goffs until later and has recently been renamed Tanlands.

STROOD FARM, PETWORTH (SU 988198). The good stone foundations have made it much more likely that this barn will survive. The smaller door was probably put in later, so that the huge doors below the porch did not have to be opened every time. William de la Strode lived here in 1328.

HAINES FARM, PETWORTH (SU 983197). An aisled barn, where local stone has been used in conjunction with some framing on the gable.

SELSCOME FARM, PETWORTH (TQ 007228). A rather neglected single storey stock shed.

RUMBOLDS FARM, PLAISTOW (TQ 002301). A wonderfully varied collection of buildings in good order, including a granary on staddle stones, set next to the farm yard pond.

Bordering on Surrey, the church was once a chapel-of-ease to Kirdford, before becoming a parish in its own right. Between 1561 and 1633 the farm was owned by the Strudwick family, who had interests in 21 other properties in the parish during the seventeenth century. They made much of their money in the local glass and iron industries.

In 1798, William Cooper of 'Rumball' had 7 cattle, 10 sheep, 4 pigs, 4 horses, 1 wagon, 2 carts, and employed a carter and a stockman.

The barn on the right is now being converted into a house.

SPRINGLANDS FARM, SHERMANBURY (TQ 229199). This was once called Taylors, and its name can still be seen painted on the back of the pew rented by the farm in Shermanbury church.

HOME FARM, SHERMANBURY (TQ 208192). The hovel set at right-angles to the barn forms the traditional 'yard' space so familiar to generations of small farmers.

MATCHETTS BARN, SHIPLEY (TQ 153227). The small door and window frame in this building suggests storage use other than hay and fodder – possibly for tools and machinery, with a raised floor for labourers' sleeping quarters.

Opposite page PUTMAN'S LANE BARN, QUEBEC, HARTING (SU 771211). A typical remote field barn.

Above GRIFFINS, WEST GRINSTEAD (TQ 181216). The history of the owners and occupants of this farm can be traced to the present day with hardly any breaks, from Griffino de Grenstede, who paid tax in 1296.

Right RIVERHILL FARM, PETWORTH (TQ 004215).
The opposing doorways suggest this was used for threshing. Coursed local stone has brick dressed air vents.

HARESFOLD FARM, WISBOROUGH GREEN (TQ 052248). The building in the foreground appears to be a long cart shed with a granary or hay loft above (note the opening in the gable) and an aisle to provide stalls for horses. The name of the farm may originate with John Hasfold, who paid 2s tax in 1332.

FAIR OAK FARM, WISTON (TQ 148143). A beautiful combination of traditional building materials. Hugh Haine was farming here when he died in 1725, leaving his widow with five underage children. There was 60lbs of cheese in the kitchen, with pork, bacon and vinegar, and among his crops were wheat, oats and barley, as well as clover, beans and cabbage. He had seventy-nine animals, including cattle, sheep, horses and pigs. His goods totalled nearly £400, at a time when £1 was equivalent to nearly £70 today.

	£ s d
Cart Harness	1.08.06
Ole Iron	5.00
Husbandry Tools	12.00
One Iron Barr Saddles & Bridles	15.06
One Grind Stone	2.06
Fewell & Chalk	3.11.09
Hog trough & Hutches	12.00
For Sowing 15 acres of Wheat & the Seeds	19.10.06
For Sowing 19 acres of Oats & the Seeds	15.10.00
For Sowing Six acres of Barley & the Seeds	7.01.00
For Three Bushells of Clover Seeds Sown	3.06.00
For One Bushell & half of Cowgrass Sown	1.16.00
Dung	4.00.00
Beans & Cabbage	6.00
Three Kilns of Lime	16.10.00
One Load of Wheat in the Straw	7.10.00
Fourteen Quarters of Oats at 12s per Quarter	8.08.00
One pair of Oxen & Six Steers	40.00.00
One pair of Oxen	15.00.00
Two Heifers & Calves	6.00.00
Three Steers & one Heifer	13.15.00
Three two Yearling Steers & two Heifers	13.15.00
Eight twelve Monthings	17.00.00
Eight Cows Seven Calves & a Bull	48.00.00
One Heifer & a Calf	3.00.00
Three Horses & one Colt	14.00.00
Six Shutts & one Sow	7.00.00
Twelve Ews & Eight Lambs	4.00.00
Husbandry Tackling	15.11.00

81

Above and opposite page HOPHURST FARM, WORTH (TQ 356385). Since this photograph was taken in 2000, the farm has been sold for conversion, and Peter Becker's collection of agricultural implements and tackle has been given to East Grinstead Museum.

The barn is a good example of a multi-purpose construction, providing stabling and a floored storage end. The interior view illustrates the kind of 'husbandry tackling' so often listed in seventeenth and eighteenth century inventories.

EASTSHAW FARM, WOOLBEDING (SU 874244). The detailing on this small building and the variety of materials used both to construct and repair, give it a value to the landscape that is difficult to measure, and could be easily destroyed by conversion of any kind. It therefore presents great problems of use and maintenance. The dentillated course at the eaves is more usual on domestic buildings, and it may have been a stable.

FARM BUILDINGS

We have seen that there are references to granaries on larger estates in thirteenth and fourteenth century documents, and archaeologists suggest that some of the traces they have found on even earlier sites are the remains of box-like granaries raised on stilts.

Probate inventories taken to go with the wills of the small farmers of the sixteenth and seventeenth centuries often refer to grain stored in the farmhouse, and it was not until the eighteenth and nineteenth centuries, with improvements in yields, that separate buildings to house grain became more common. These stores could either be at first-floor over a cart-shed, or in a purpose-built

MITFORD HOME FARM. Henry Hooker with his turkeys in 1931. 2¼ cwt grain sacks had to be man-handled up the granary steps, although some such buildings incorporated hoists. The ground floor was used as a cart shed.

structure raised on staddle-stones or brick piers as protection against damp and vermin. Most of those that now survive date from the eighteenth or nineteenth century, and generally their survival has depended on how well they can be adapted to a different use.

Horse-breeding and training has become widespread in West Sussex in recent decades, and appropriately the county is home to a racecourse at Goodwood and the showground of Hickstead. Place names indicate that this has a long pedigree. *Horse*-ham, *Warn*-ham and *Sted*-ham were settlements important for keeping or breeding horses, mares and stallions. In earlier centuries their chief purpose was as a means of transport and as draught animals, although in the Weald oxen were more widely used, as they were better able to cope with the heavy soils.

Writing in the early thirteenth century, the steward to the bishop of Chichester gave the prices of a palfrey, a war-horse, cart horses, plough horses and pack horses, and recommended that 'if more carts are advisable, 12 mares should be borrowed for them as horses sell as dear as gold in Sussex'.

By the time the Rev Arthur Young made his surveys at the end of the eighteenth century, horses were in general use, and he felt that the local animals were of high quality, although they had still not displaced oxen completely.

Many of the larger farmyard complexes included a small smithy, which could cope with repairs and modification to tools and machinery, as well as shoeing horses. Even this was nothing new, as the thirteenth century smith at Amberley had to mend ploughs, shoe horses for the bishop and the estate officials, and grind the harvest scythes and sheep shears as part of the obligations on the 4 or 5 acres that went with his smithy. In later periods, it was the smith who would manufacture the metal straps so often used to repair framed buildings.

The Romans bred pigeons for food in specialist buildings, but the earliest surviving dovecotes are no earlier than the twelfth century. Originally this was a privilege solely of the lords of manors, as a valuable source of fresh meat and eggs throughout the year, but this was challenged in the seventeenth century with the rise of the gentleman farmer, and dovecots became more widespread. The heyday of the dovecot was the period 1640 to 1750 when grain was cheap. Although there are many examples, some dated, of large stone dovecots associated with the larger houses, like that at Trotton, there is also evidence for small cotes integrated into the ends of barns, as at Okehurst near Billingshurst.

MANOR FARM, EASTERGATE (SU 945051). Fine two-storey framed granary, of the sixteenth century, as is the timber-framed farmhouse.

Granaries

Opposite page CHURCH FARM, ITCHENOR (SU 801007). Pyramidal roof to a square building.

Above WOOLHURST FARM, LODSWORTH (SU 925231). Vertical and horizontal boarding.

Opposite page MANOR FARM, WEST DEAN (SU 865126). A unique hexagonal solid-walled granary on staddles, and still with a thatched roof.

Above TRENCHMORE FARM, LOXWOOD (TQ 057299). A boarded building on brick piers; the parish of Loxwood was originally a chapel-of-ease to Wisborough Green. Alan, the captain of one of Richard I's ships, called 'The Sea-Plough' *(Trenche-la-mer)* took his name from the vessel he commanded – 'Trenche-mer'. It was probably his son who was in charge of royal galleys in 1217, and who gave land and a saltpan near his home in Shoreham to the Knights Templar, in 1220. By the 1300s, men called Trenchmore were living in the Weald.

DIDDLESFOLD MANOR, NORTHCHAPEL (SU 949295). An example of the use of brick piers to replace some of the staddle stones.

NORTHOVER FARM, STEYNING (TQ 181148). Cart shed with granary over. In about 1820, this farm on the Wiston estate consisted of 220 acres.

LODGE FARM, TILLINGTON (SU 941243). Now converted to a 'business centre', the farmhouse has long since gone. What does survive – if only just – is this boarded granary on staddle stones, which still has its grain bins.

BAILING HILL FARM, WARNHAM (TQ 153328). Timber-framed granary over cart shed.

POPLARS PLACE, WORTH (TQ 336387). Boarded granary on staddle stones.

Stables

FITTLEWORTH HOUSE FARM, FITTLEWORTH (TQ 009196). Built in local stone, the clue that this is not merely another open-sided cattle hovel lies in the chimney stack. This probably served a small farmyard forge.

Opposite page top GOWNFOLD FARM, KIRDFORD (TQ 012258). Another single-storey stone building that almost certainly contained a forge used for smithing and possibly repairs to carts.

Opposite page below NEW BARN FARM, MADEHURST (SU 982103). Two ranges of single-storey buildings commonly used as stabling, although they may have started life as milking sheds

Above CLAPHAM FARM, CLAPHAM (TQ 095066). Farm buildings are flexible, and can sometimes be adapted to a variety of uses. Although this flint and brick building looks like many small forges, the pig pens beside it may indicate an alternative use as a slaughter shed. It was originally a dovecote and has three feet thick walls.

BERRYWOOD FARM, HEYSHOTT (SU 895183). A barn containing stabling and hay storage, in a mixture of stone, brick and boarded framework, typical of present-day equestrian establishments.

Other Buildings

HOLT PLACE FARM, BIRDHAM (SZ 811993). A clock tower over low stock sheds could be a sign that agriculture was moving away from being part of the fabric of life towards becoming an industry like any other, obsessed with time-keeping.

Above DIDLING MANOR FARM, TREYFORD (SU 838186). Although flint barns often have brick dressed vents, this cluster of brick framed openings suggests that it may be associated with a manorial dovecote, incorporated into the barn.

Left SADDLESCOMBE FARM, NEWTIMBER (TQ 278115). This attempt to provide even the smallest hamlet with modern facilities is in marked contrast with the way in which villages today are losing local services.

Opposite page TRULEIGH MANOR FARM, UPPER BEEDING (TQ 225115). A formal entry into a planned farmyard has offered the opportunity for a purpose built dovecote over the archway. The castellations and cone-shaped gate-post were added by Captain Masters in the 1920s, who as well as farming was Master of Foxhounds and well-known local eccentric.

Opposite page HARWOODS GREEN FARM, STOPHAM (TQ 032205) and *Above* CLAPHAM FARM, CLAPHAM (TQ 095066).

These two interiors are evocative of the ways farm barns could be used for all manner of purposes, most of which have now been made obsolete by modern machinery, storage using metal framing, sheeting and plastic and industrial farming practices.

CLAPHAM FARM, CLAPHAM (TQ 095066). Curiously this chimney stack is purely decorative, and has never served a hearth or furnace in the memory of the farmer. It is of the same design as the chimneys on the farmhouse buildings, but is a real 'folly'! (*see also the previous page*).

DECAY AND RESTORATION

Allowing for brief periods of prosperity during the two World Wars, traditional farming has been in decline from the beginning of the twentieth century, and so have its buildings. Increased mechanisation and industrialisation has made traditional barns, granaries and cattle hovels redundant or impractical. Dutch barns demonstrated that hay storage did not need buildings with walls, and old barns could not accomodate ever-larger machinery. Sadly, we have become accustomed to seeing barns decaying into the ground, especially those that are more remote, and being replaced by unsightly modern versions in metal, or stacks of plastic wrapped bales.

As long as buildings were as much a declaration of status as working plant, local estate owners, even small farmers, were once proud to mark their contributions with date stones, and to add decorative but non-functional details. Modern large-scale commercial farming operations no longer have loyalties to an estate or a locality and their buildings are uniform across the country, utilitarian and purely functional, with far less emphasis on permanence.

In the 1960s and '70s concern was turned into action, and those decades saw increasing numbers of 'barn conversions', mainly for residential use. Unfortunately, this anxiety over the potential loss to the history of the region and the rural landscape produced rather hasty and sometimes very unsuitable results. In the face of a welter of applications, planning authorities have brought out guidance notes to encourage restoration that is best suited to both the buildings and their surroundings.

Nowadays, before agreeing to a house conversion, every effort is made to find a use that will keep the original character and form of the building, and if possible benefit the local community in some way. Converting a redundant farm building into a house is perhaps one of the least

suitable options and can be the most damaging to the fabric of the building. This was not so in earlier times when the difference between residential and farm buildings was not so great. Recently, at Cowfold in the Weald, a redundant farm yard with twentieth century brick buildings was sold for residential development. The old farmhouse had been demolished and replaced in the 1960s, leaving one dilapidated timber barn. The discovery that this barn had a crown-posted roof led to the uncovering of the complete history of the site.

A reference to the farm, with a farm house, appeared in a document of 1373. In the manorial records for 1583, the tenant was granted permission to demolish the existing house and rebuild. It became clear that he did rebuild – about 200 yards away – in the latest style with chimneys, but that he did not demolish the old house. He downgraded it into a barn, and a barn it has been ever since, with a single bay added in the eighteenth century. The 'new' farmhouse was then demolished in the 1960s. The 'barn' is now in the process of being restored to a house.

There seems to be the evidence for a similar story in the village of Bury, just north of the Downs, where there is a barn with a number of residential characteristics, and the house close by that was probably a 'modern' replacement in the seventeenth century, contains medieval joists that may have come from the 'barn'. Unfortunately the expectations of modern living create greater demands for structural change.

The problems of restoring buildings other than barns and still keeping their character are even greater. Stables have been brought back into use with the increased popularity of equestrian sports, but what to do with small granaries, forges, cattle hovels, cart-sheds? Ideally they should be kept in their original setting, but finding viable

uses when their agricultural life has finished can be difficult, and they are often far more costly to repair and maintain than modern equivalents. It requires commitment and vision from owners, and fortunately there are almost as many of these as profit-driven developers cashing in on public nostalgia.

SLINFOLD FARM, SLINFOLD. A house contemporary with many of the farm buildings illustrated, showing the similarity between the structures.

Opposite page PEACOCKS FARM, NORTHCHAPEL (SU 962294). Fifteen years separate these two pictures, the last taken in 2000, and they illustrate the way in which traditional farm buildings are lost, as they decay into the ground leaving little trace. Its position by the roadside suggests it was once a cart shed, but the will is needed to find new uses for such buildings. It was finally demolished in 2002.

Above ANDREWS FARM, WARNHAM (TQ 165345). Traditional boarded buildings cheek by jowl with a modern Dutch barn (left) and a metal storage tank (right).

Opposite page CHAITES FARM, BOLNEY (TQ 277216). A typical example of the sad cycle of neglect and decay, once the protective weather-boarding is no longer maintained.

Opposite page top MILL FARM, LURGASHALL (SU 936259). Once a fine group of buildings, with barn and cattle shelters, now fast being overtaken by brambles and elder.

Opposite page below HARWOODS GREEN FARM, STOPHAM (TQ 032205). This barn, in a mixture of stone and timber with half-hipped tiled roof, has been repaired in the past with a stone buttress. The present state of the barn suggests there is no longer an interest in repair work.

Above COLEHOOK FARM, EBERNOE (SU 968277). Another remote barn with an end lean-to once used for cattle, now falling into decay.

Above CHURCH FARM, CHIDHAM (SU 788039).
Opposite page top THE DOVECOTE, EDBURTON (UPPER BEEDING) (TQ 235115).
Opposite page below KENT BARN, HARTING (SU 795197).

The conversion of farm buildings to dwellings can illustrate the conflicts between very different demands and expectations – how to convert a piece of historical agricultural plant into a house without losing the essential character, in an age when there is no longer a basic similarity between the two kinds of structure.

These three conversions have ensured survival while still reflecting local materials, and traditional proportions. All three have glazed the original large door openings – although none have retained the barn doors, which can act as 'shutters' by night, improving both heating and security.

The Dovecote is a conversion very close to Edburton church, and has preserved an important group, although the particular character of the manorial dovecote has been somewhat lost. All have avoided the intrusive dormer windows and porches that can obscure the original agricultural function of such buildings, and for the same reason have opted for simple flues rather than chimney stacks.

FURTHER READING

Bruce, P., *Northchapel, A Parish History* (Northchapel Parish Council), 2000
Hughes, G., *Barns of Rural Britain* (Herbert), 1985
Jerome, P., *All is not sunshine hear* (The Window Press), 1996
Sleight, J. (ed), *Yeoman Farmers and Gentlemen; People of Wiston, West Sussex, 1612-1732* (CCE Sussex), 1993
Wade, Martin. S., *Historic Farm Buildings* (Batsford), 1991
Warren, J. (ed), *Wealden Buildings (chs 9, 10, 11)* (Coach Publishing), 1990
Wiston Estate Study Group, *'All is safely gathered in'* (WSCC), 1990
Young, Rev. A., *General view of the Agriculture of the County of Sussex, 1813* (David & Charles), reprint 1970

ARTICLES

Cornwall, J., *Farming in Sussex 1560-1640* (Sussex Archaeological Collections vol 92.48-92), 1954
Agricultural Improvement 1560-1640 (Sussex Archaeological Collections vol 98.118-132), 1960
Farrant, J.H., *'Spirited and intelligent farmers': The Arthur Youngs and the Board of Agriculture's reports on Sussex, 1793 and 1808* (Sussex Archaeological Collections vol 130.200-212), 1992
Gardiner, M., *The Medieval Rural Economy and Landscape in An Historical Atlas of Sussex* (Phillimore), 1999
Godman, P.S. & Hudson, W., *On a Series of Rolls of the Manor of Wiston* (Sussex Archaeological Collections vol 54.130-182), 1911
Hudson, W., *On a Series of Rolls of the Manor of Wiston* (Sussex Archaeological Collections vol 53.143-187), 1910
Kenyon, H., *Kirdford Inventories 1611-1776* (Sussex Archaeological Collections vol 93.78-156), 1955
The Civil Defence and Livestock Returns for Sussex in 1801 (Sussex Archaeological Collections vol 89.57-84), 1950
Salzman, L.F. (ed), *Ministers' Accounts of the Manor of Petworth 1347-53* (Sussex Record Society vol 55), 1955
Short, B., *Agricultural Regions, Improvements and Land Use c1840 in An Historical Atlas of Sussex* (Phillimore), 1999
Wilson, A.E., *Farming in Sussex in the Middle Ages* (Sussex Archaeological Collections vol 97.98-119), 1959

PHOTOGRAPHIC ACKNOWLEDGEMENTS

Without the kindness and co-operation of the following farmers, landowners and estates this book would not have been possible. My sincere thanks to them all; I am much indebted.

Mrs. Julia Baker; Balcombe Estate; Mrs. K.H.C. Bowyer; Mr. Peter Becker; Mr. Duncan Branch; Mr. David Burden; Mr. Mick Childs; Mr. Andrew Collins; Eileen, David and Peter Cornford; Cowdray Estate; Mr. Brian Dallyn; Mrs. E.A. Ellson; Mr. and Mrs. J. Garrett; Mr. Andrew Gibb; Mr. J. Goring; Mrs. R.I Grant; Mr. R.G.L Haydon; Mr. J. Helyer; Mr. lan Hughes; Mr. J.D. Kempley; Mr. Graham Kittle; Knepp Castle Estate; Leconfield Estate; Mr. & Mrs. Mark Lee; Mr. E. Lock; Mr. Peter Lovejoy; Mr. Jonathon Lucas, Mr. R.J. Lywood; Mr. D. Mortimer, The National Trust; Mrs. P.A. and the late Mr. William Payne; Mr. Andrew Donald Pick; Mr. Robert Scott; Mr. John Seward, Basil and Paul Strudwick; Mr. John Trenchard; West Dean Estate; Robin and Sandra Windus; Wiston Estate; Mr. David Wright.

I am also grateful to Charles Skey for his help and encouragement, and Jumbo Taylor and Robin Upton for all their help. Christopher Chaplin kindly drew the map and all the diagrams.

I would also like to thank the following for allowing the inclusion of black and white photographs which are either in their possession or for which they hold the copyright: Mr. Ron Barber (Grittenham Farm, page 12), Mr. C.W. Cramp (Slinfold Farm, page 108), Mr. Martin Hayes and Mr. Robin Knibb of Worthing Library (Lagness Farm, page 19), Mr. David W. Morris (The Old Barn, Old Place, page 20), Mrs. Gena Wilmhurst of the Storrington Musuem (Green Farm, page 6), and West Sussex Record Office, © The Garland Collection (Mitford Home Farm, page 85).

The photograph of a farmyard near Chanctonbury on page 4 is taken from *English Downland* by H.J. Massingham (Batsford 1936) and the view of Fulking High Street on page 10 comes from *A Chronicle of Edburton and Fulking, Sussex* (Hubners Ltd, 1958).

Special thanks are owed to my wife Sue, who for ten years has accompanied me on virtually all of my photographic expeditions in search of barns in West Sussex, and whose constant support, patience and tolerance have helped make this project so enjoyable.

INDEX

Adur, river 7
Agrarian Revolution 11
Albourne, Inholmes Farm 2 & 3
Amberley 10, 25, 26, 86
Amberley, Castle Farm 7, 8, 18
Ardingly (Balcombe), Ryelands Farm 54
Arun, river 7
Arundel 10, 31
    Castle 25
    estate 9, 25, 50
    FitzAlans of 60
Ashington 10
ate Ree, John 54

Bartelot family 43
Becker, Peter 82
Bedham manor 58
Billingshurst, Rowner Farm, 55
Birdham 10
    Holt Place Farm 101
Black Death 11
Blackehurst, Warin & Eloise of 50
    William of 50
Blann, Michael 21
Bolney, Chaites Farm 110, 111
    Garstons Farm, The Granary 52, 53
    Nyes Hill Farm 56
Bone Barn, 32
    Field 32
Boxall, John 20
Bramber 10
    Castle 26
Brighton 27
Broadbridge Heath 45
Brown, Lilian 32
Brownes, the 12
Burpham, Burpham Farm 19
Burrells, the 13
Bury 107
Bury 25, 27
    (near Bignor) 22

Manor Farm 31
    Southview Farm 32

Canterbury, Archbishop of 25, 26, 40
Carylls, the 12
Chanctonbury 4
    Ring 22
Chichester, Bishop of 9, 11, 12, 25, 26, 28, 86
Chidham, Church Farm 114
Clapham 25
    Clapham Farm 98, 99, 105, 106
    Clapham Manor Farm 33
    Longfurlong Barn 34
Coldwaltham, Besley Farm 34
Coleman, Richard 17
Compton, Watergate House Farm 35
Constable, John 13
Cooper, William 72
Cowdray, Brownes of 25, 38
Cowfold 107
Cox, Thomas 62

de Braose family 27
de Covert, Baldwin 45
de Grenstede, Griffino 78
de l'Isle family 20
de la Strode, William 69
Domesday Survey 10
Downer family 58
Downs, the 9, 11, 22, 107

Easebourne 12
East Grinstead Museum 82
Eastergate, Manor Farm 86, 87
Ebernoe, Colehook Farm 113
    Old Elkham Farm 60, 61
Edburton (Upper Beeding), The Dovecote 114, 115
Elkham, John 60

Ferring estate 26, 27
Ffyst, Robert 60
Fittleworth, Fittleworth House Farm 97
Fulking (Edburton parish) 10

Garston family 53
Goodwood 86
    estate, Dukes of Richmond 25
Goring family 12, 22, 55
Graffham, Smokey House Barn 36
Great Storm 1987 22

Haine, Hugh 81
Hardham 12
Harting, Kent Barn 114, 115
    Quebec, Putman's Lane Barn 76, 77
Hastings 10
Henry VIII 12, 25
Heyshott, Berrywood Farm 100
    Legges Barn 37
Hickstead 86
Hooker, Henry 85
Horsham 26
Hussey family 20
Ifield 10

Itchenor, Church Farm, 88, 89
Itchingfield, Possession House Farm 57

Kilhams, Charles 31
Kirdford, Crimborne House Stud Farm 58, 59
    Gownfold Farm 98, 99
    Marshalls Farm 20, 58
Knight, William 58
Knights Templar 38, 91
Knights Templars 23, 27

Lavant 10
    Little Oldwick Farm 29
le Honte, John 26

118

Lewes 10
Lindfield, Scaynes Hill, Butterbox Farm 17
    Scaynes Hill, Sennotts Farm 15
Littlehampton 7
Lodsworth, Godsdenheath Farm 64, 65
    Woolhurst Farm 65, 89
Lower Chancton Farm 22
Loxwood, Trenchmore Farm 91
Lurgashall, Mill Farm 112, 113

Madehurst, New Barn Farm 99
Magicke, John 68
Mareshal, Thomas 58
Markham, Gervase 12
Mascall, Leonard 12
Masters, Captain 102
Michelgrove 13, 33
Midhurst, Cowdray estate 25
Mitford Home Farm 85
Moss, Algie 65

Napoleonic Wars 13
National Trust 38, 40
Newtimber, Saddlescombe Farm 23, 38, 39,
    102
Norden, John 12
Norfolk, 12th Duke of 33
    Duke of 27
    Dukes of 31
    Dukes of, Fitzalan and Howard 25
Norman Conquest 9, 11
Northchapel, Diddlesfold Manor 92
    Ebernoe, Frithfold Farm 62, 63
    Goffs Farm 68
    Peacocks Farm 108, 109

Okehurst 86

Pagham, Lagness Farm 19
Pallingham manor 60
Parham, Bishops of 12
    Palmers of 50
Pasmore, Mrs 45
Petworth 25
    estate 23, 58
    Leconfields 25
    Percy family 25
    Haines Farm 70
    Percies of 27, 30

Selscome Farm 71
Stag Park Farm 13
Strood Farm 69
Riverhill Farm 78, 79
Westlands Farm 65, 66, 67
Pevensey 10
Plaistow, Rumbolds Farm 72, 73
Poynings, Wickhurst Barn 40
Pulborough 27
    Old Place, The Old Barn 20

Richard I 91
Robinson family 38
Rowner family 55
Rudgwick 26
Rusper 12

Saddlescombe 25, 27
Sele Priory 26
Selsey 7
Shelley family 13, 33
Shermanbury, Home Farm 75
    Springlands Farm 74
Shipley 27
    Matchetts Barn 76
    Pondtail Farm 11
Shoreham 7, 91
Sidlesham, Church Farm 28
Slindon, estate 26
    Warren Barn 41
Slinfold, Slinfold Farm 108
South Stoke, South Stoke Farm 42
Steyning 27
    Northover Farm 93
Stopham, Harwoods Green Farm 104, 105,
    112, 113
    Turnpike Farm 43
Storrington 10
Strudwick family 72
    Basil 9
Sullington, Manor Farm 16, 45, 47
    The Great Barn 46, 47
Syon Abbey 12

Tanlands 68
Taylor, William 68
The Sea Plough 91
Tillington, Grittenham Farm 12
    Lodge Farm 94

Tompkins 34
    family 33
Trenche-mer, Alan 91
Treyford, Didling Manor Farm 48, 49, 50,
    102
Trotton 86

Upper Beeding, Truleigh Manor Farm 102,
    103
    Upper Horton Farm 21
Upwaltham, Upwaltham Manor Farm 14

Walberton, Binstead Nursery Farm 29
    Manor Farm 44
Warnham 13
    Andrews Farm 110
    Bailing Hill Farm 95
Warningcamp 25
    Blakehurst Farm 50
Washington, Church Farm 24
    East Clayton Farm 51
    Gerrard 24
    Green Farm 6
    Rock Mill 6
Weald, the 9, 12, 13, 14, 22, 26, 27, 52, 86,
    91, 107
West Dean, Manor Farm 90, 91
West Grinstead, Griffins 78
    Pondtail Farm 13
Westminster Abbey 12
William the Conqueror 10
Wisborough Green 91
    Haresfold Farm 80
Wiston 27
    estate 22, 60, 93
    Fair Oak Farm 81
    Wiston Great Barn 18
Woolbeding, Eastshaw Farm 84
Worth, Hophurst Farm 82, 83
    Poplars Place 96
Worthing Museum 21

Yaldwin, William 68
Yapton, Bonhams Farm 30
Young, Rev Arthur 86

Zouche family 30

119

LIST OF SUBSCRIBERS

The publishers would like to thank all those whose names are listed below,
as well as the many subscribers who chose to remain anonymous.
Their support, and interest in the farm buildings of West Sussex,
helped make this book possible

Caroline Adams
Nona Almond
B. Almond
Peter and Beryl Anderson
G.H. Armstrong
William C. Aslett
S. Atterton

Keith & Karen Baker
Brian Banister
Robin H. Barnes
Peter C. Benner
Stella Berrisford
Roy Berry
Michael Bevan
Sheila Blair
David, Anne & Jessica Bone
John Boxall
Mary Boxall
C.W.B. Boyer
J.M. Bray
E.G. Brice Esq. J.P.
Iain A. Brown
Mrs. Judith E. Brown
Richard J. Bryant
Victor Robert Burns

John C. Charman
Diana Chatwin

Mr. & Mrs. T.C. Churcher
Victoria & Imani Clare
Jane P. le Cluse
Mr. & Mrs. Richard and Marian Coles
Mr. Stuart Coles
P.D. Combes
John Cooper
David Boaz Cornford
Dr. Peter Corrigan
Nigel D. Courtnage
K. Coutin

Professor R.P. Dales
Robert Geoffrey Dancy
Brian Dawson
Lucy Dewhurst
J.H. Dicks
Nigel F. Divers
Susan C. Djabri
S. Dunford

David and Mieke Earl
Mrs. Joan M. Earl
Adrian and Kate Earl
Toby & Juliet Earl-Jones
David W. Meryon Easton
Eric Ralph Edmunds
J. Michael Edwards
Mrs. Isobel Evans

O. & R. Farley
Ian Farrell
The Fernhurst Society
Ferring History Group
Margaret A. Fisher

Richard Garne
Elizabeth A.S. Garrett
A.M. Gillespie
Mr. David Goodger
Mary Greenfield
Walter Greenway
Christopher Grinsted

Michael Hall
Douglass & Susan Hallock
Miss E.M.S. Hammond
Mr. & Mrs. M.D. Harding
R.G.L. Haydon
Elsa and Jim Haynes
Leslie R. Hobbs
E.A. Hollingdale
Barry Homewood
Horsham Museum Society
Theo & Sammantha Hughes
R.F. Hunnisett
Mr. & Mrs. A.J. Huntley
John Hurd
Peter Iden

Peter James
Daniel Johnston
R.L. & P.A. Jones

Gordon Kearvell
Ann Knowles

Dr. Patrick A. Leadbeater
Frank Leeson
Kim Leslie
G. & E. Lloyd
Garry Long
Gordon Long
Peter Longley

Dr. Philip MacDougall
Freddie Madgwick
Mr. & Mrs. T. Marsden
A.E. Marshall
David Marshall
Brian Maystone
Valerie A. Mellor
Rod Mengham
Wendy Millar
Alan J. Mitchell
John Morrish

Bob and Miranda Northover
Brian and Daphne Norton

Celia Parrott
David & Joan Pateman
Peter James Payne
Norman Peachell

John Pelling
Janet Pennington
Shirley Penny
John and Errol Phillipson
David and Helen Philpott
Mrs. Pamela Platt
R.J.A. Plummer
Nigel Purchase

David M. Raven
Norman Read
Colin E. Reid
J. Spencer Richards
Ian John Richardson
Karen Richardson
Shirley Ridley
Robert William Robb
Don Roberts
T.H. Rose
Robbin Rouse
Mr. Stephen Russell

Peter J. Saunders
Marie Schlotter
Mr. S.C. Selby
Geoffrey Sheard
Mrs. A. Sherman
Professor Brian Short
Sheila & Susan Silver
Dennis Sivyer
Justin and Marie Smith
Paul Smith
Alan D. Soutter
Barry Sowton

Peter J. Squelch
Nick Steer
M. Suckling
Sussex Downs Conservation Board
Mrs. B.C. Symington

L.J. Taylor
Jeff & Rosalind Thompson
Heather Thomson
K.M. Tong
Sean Trendall
Ann Turner

Percy and Daphne Upton

Michael Verrall

Heather Warne
Professor Brian Warner
Paul & Samantha Waters
Mrs. Phyl Webb
Mrs. Norma Weir
Leslie Weller
Deborah Wells
Mike West
P.W.C. White
Tom White
Mary Wigan
Kathy Wigan
Maev Wilkinson
Mr. Robin Windus
Worthing Archaeological Society